対訳

ISO 31000:2018

(JIS Q 31000:2019)

ポケット版

リスクマネジメントの国際規格

日本規格協会　編

ISO copyright office
CP 401・Ch. de Blandonnet 8
CH-1214 Vernier, Geneva
Phone: +41 22 749 01 11
Fax: +41 22 749 09 47
Email: copyright@iso.org
Website: www.iso.org

Published in Switzerland

本書について

本書は，国際標準化機構（ISO）が 2018 年 2 月に第 2 版として発行した国際規格 ISO 31000:2018 (Risk management − Guidelines)，及びその翻訳規格として，日本工業標準調査会（JISC）の審議を経て 2019 年 1 月 21 日に経済産業大臣が改正した日本工業規格 JIS Q 31000:2019（リスクマネジメント－指針）を，英和対訳で収録したものです．

収録に際して JIS の解説は省略しています．JIS の解説を参照したい場合は，JIS 規格票をご利用ください．

また，規格をより深く理解し実践したい方には，日本規格協会より発行予定の解説書籍を併読されることをお勧めします．

2019 年 3 月

日本規格協会

Contents

ISO 31000:2018
Risk management — Guidelines

目　次

JIS Q 31000:2019
リスクマネジメント－指針

Foreword

ISO (the International Organization for Standardization) is a worldwide federation of national standards bodies (ISO member bodies). The work of preparing International Standards is normally carried out through ISO technical committees. Each member body interested in a subject for which a technical committee has been established has the right to be represented on that committee. International organizations, governmental and non-governmental, in liaison with ISO, also take part in the work. ISO collaborates closely with the International Electrotechnical Commission (IEC) on all matters of electrotechnical standardization.

The procedures used to develop this document and those intended for its further maintenance are described in the ISO/IEC Directives, Part 1. In particular the different approval criteria needed for the different types of ISO documents should be noted. This document was drafted in accordance

まえがき

(ISO の Foreword と JIS のまえがきは，それぞれの原文において内容が異なっているため，対訳となっていないことにご注意ください.)

この規格は，工業標準化法に基づき，日本工業標準調査会の審議を経て，経済産業大臣が改正した日本工業規格である．これによって，**JIS Q 31000**:2010 は改正され，この規格に置き換えられた．

この規格は，著作権法で保護対象となっている著作物である．

この規格の一部が，特許権，出願公開後の特許出願又は実用新案権に抵触する可能性があることに注意を喚起する．経済産業大臣及び日本工業標準調査会は，このような特許権，出願公開後の特許出願及び実用新案権に関わる確認について，責任はもたない．

with the editorial rules of the ISO/IEC Directives, Part 2 (see **www.iso.org/directives**).

Attention is drawn to the possibility that some of the elements of this document may be the subject of patent rights. ISO shall not be held responsible for identifying any or all such patent rights. Details of any patent rights identified during the development of the document will be in the Introduction and/or on the ISO list of patent declarations received (see **www.iso.org/patents**).

Any trade name used in this document is information given for the convenience of users and does not constitute an endorsement.

For an explanation on the voluntary nature of standards, the meaning of ISO specific terms and expressions related to conformity assessment, as well as information about ISO's adherence to the World Trade Organization (WTO) principles in the Technical Barriers to Trade (TBT) see the following URL: **www.iso.org/iso/foreword.html**.

This document was prepared by Technical Committee ISO/TC 262, *Risk management.*

This second edition cancels and replaces the first edition (ISO 31000:2009) which has been technically revised.

The main changes compared to the previous edition are as follows:

— review of the principles of risk management, which are the key criteria for its success;

— highlighting of the leadership by top management and the integration of risk management, starting with the governance of the organization;

— greater emphasis on the iterative nature of risk management, noting that new experiences, knowledge and analysis can lead to a revision of process elements, actions and controls at each stage of the process;

— streamlining of the content with greater focus on sustaining an open systems model to fit multiple needs and contexts.

Introduction

This document is for use by people who create and protect value in organizations by managing risks, making decisions, setting and achieving objectives and improving performance.

Organizations of all types and sizes face external and internal factors and influences that make it uncertain whether they will achieve their objectives.

Managing risk is iterative and assists organizations in setting strategy, achieving objectives and making informed decisions.

Managing risk is part of governance and leadership, and is fundamental to how the organization is managed at all levels. It contributes to the im-

序文

（ISO の Introduction と JIS の序文は，それぞれの原文において内容が異なっているため，対訳となっていないことにご注意ください.）

この規格は，2018 年に第 2 版として発行された **ISO 31000** を基に，技術的内容及び構成を変更することなく作成した日本工業規格である.

この規格は，リスクのマネジメントを行い，意思を決定し，目的の設定及び達成を行い，並びにパフォーマンスの改善のために，組織における価値を創造し保護する人々が使用するためのものである.

あらゆる業態及び規模の組織は，自らの目的達成の成否を不確かにする外部及び内部の要素並びに影響力に直面している.

リスクマネジメントは，反復して行うものであり，戦略の決定，目的の達成及び十分な情報に基づいた決定に当たって組織を支援する.

リスクマネジメントは，組織統治及びリーダーシップの一部であり，あらゆるレベルで組織のマネジメントを行うことの基礎となる. リスクマネジメン

provement of management systems.

Managing risk is part of all activities associated with an organization and includes interaction with stakeholders.

Managing risk considers the external and internal context of the organization, including human behaviour and cultural factors.

Managing risk is based on the principles, framework and process outlined in this document, as illustrated in **Figure 1**. These components might already exist in full or in part within the organization, however, they might need to be adapted or improved so that managing risk is efficient, effective and consistent.

トは，マネジメントシステムの改善に寄与する．

リスクマネジメントは，組織に関連する全ての活動の一部であり，ステークホルダとのやり取りを含む．

リスクマジメントは，人間の行動及び文化的要素を含めた組織の外部及び内部の状況を考慮するものである．

リスクマネジメントは，**図 1** に示すように，この規格に記載する原則，枠組み及びプロセスに基づいて行われる．これらの構成要素は，組織の中にその全て又は一部が既に存在することもあるが，リスクマネジメントが効率的に，効果的に，かつ，一貫性をもって行われるようにするためには，それらを適応又は改善する必要がある場合もある．

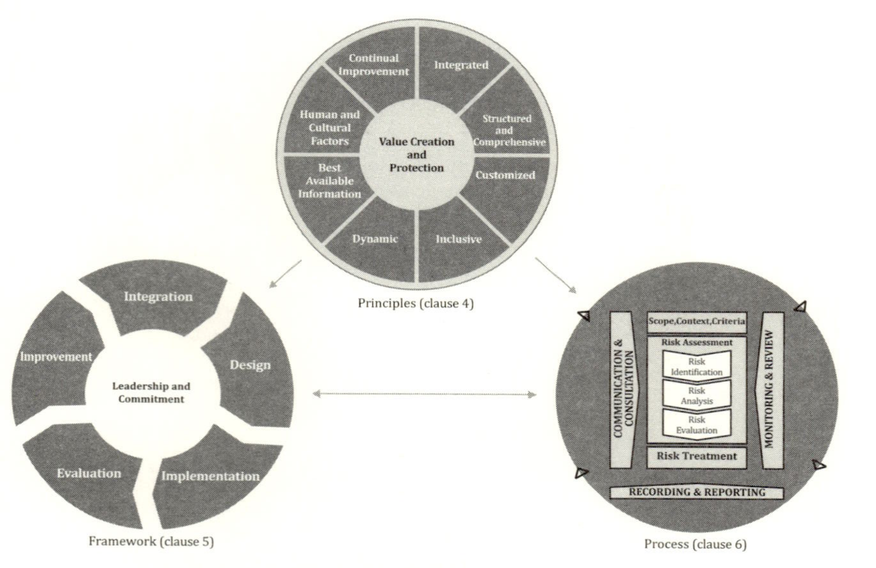

Figure 1 — Principles, framework and process

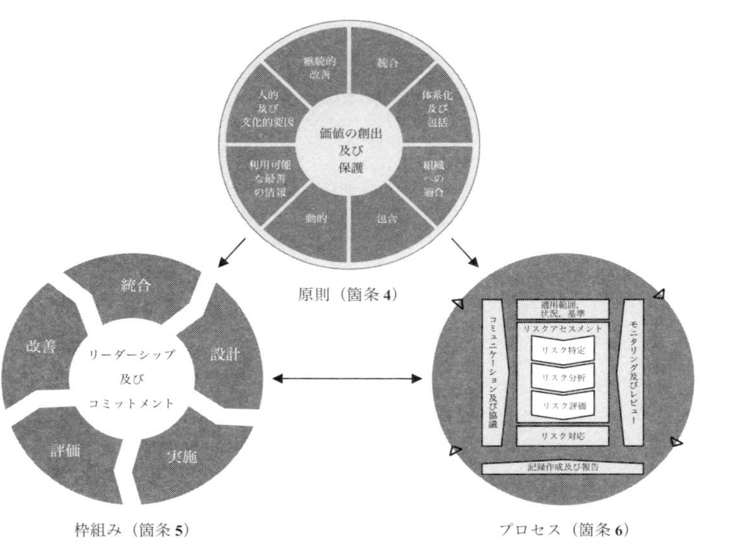

図 1 －原則，枠組み及びプロセス

1 Scope

This document provides guidelines on managing risk faced by organizations. The application of these guidelines can be customized to any organization and its context.

This document provides a common approach to managing any type of risk and is not industry or sector specific.

This document can be used throughout the life of the organization and can be applied to any activity, including decision-making at all levels.

2 Normative references

There are no normative references in this docu-

1　適用範囲

この規格は，組織が直面するリスクのマネジメントを行うことに関して，適用可能な指針を示す．これらの指針は，あらゆる組織及びその状況に合わせて適用することができる．

この規格は，あらゆる種類のリスクのマネジメントを行うための共通の取組み方を提供しており，特定の産業又は部門に限るものではない．

この規格は，組織が存在している限り使用可能であり，あらゆるレベルにおける意思決定を含め，全ての活動に適用できる．

注記　この規格の対応国際規格及びその対応の程度を表す記号を，次に示す．

ISO 31000:2018，Risk management − Guidelines (IDT)

なお，対応の程度を表す記号“IDT”は，**ISO/IEC Guide 21-1** に基づき，“一致している”ことを示す．

2　引用規格

この規格には，引用規格はない．

ment.

3 Terms and definitions

For the purposes of this document, the following terms and definitions apply.

ISO and IEC maintain terminological databases for use in standardization at the following addresses:

— ISO Online browsing platform: available at **http://www.iso.org/obp**

— IEC Electropedia: available at **http://www.electropedia.org**

3.1

risk

effect of uncertainty on objectives

Note 1 to entry: An effect is a deviation from the expected. It can be positive, negative or both, and can address, create or result in opportunities and threats.

3 用語及び定義

この規格で用いる主な用語及び定義は，次による．

なお，**ISO** 及び **IEC** は，標準化に使用するための用語上のデータベースを次のアドレスに維持している．

— ISO Online browsing platform：http://www.iso.org/obp

— IEC Electropedia：http://www.electropedia.org/

3.1

リスク（risk）

目的に対する不確かさの影響．

注記 1 影響とは，期待されていることからかい（乖）離することをいう．影響には，好ましいもの，好ましくないもの，又はその両方の場合があり得る．影響は，機会又は脅威を示したり，創り出したり，もたらしたりすることが

Note 2 to entry: Objectives can have different aspects and categories, and can be applied at different levels.
Note 3 to entry: Risk is usually expressed in terms of *risk sources* (**3.4**), potential *events* (**3.5**), their *consequences* (**3.6**) and their *likelihood* (**3.7**).

3.2

risk management

coordinated activities to direct and control an organization with regard to *risk* (**3.1**)

3.3

stakeholder

person or organization that can affect, be affected by, or perceive themselves to be affected by a decision or activity

Note 1 to entry: The term "interested party" can be used as an alternative to "stakeholder".

3.4

あり得る．

注記 2　目的は，様々な側面及び分野をもつことがある．また，様々なレベルで適用されることがある．

注記 3　一般に，リスクは，リスク源（**3.4**），起こり得る事象（**3.5**）及びそれらの結果（**3.6**）並びに起こりやすさ（**3.7**）として表される．

3.2

リスクマネジメント（risk management）

リスク（**3.1**）について，組織を指揮統制するための調整された活動．

3.3

ステークホルダ（stakeholder）

ある決定事項若しくは活動に影響を与え得るか，その影響を受け得るか又はその影響を受けると認識している，個人又は組織．

注記　“利害関係者”を“ステークホルダ”の代わりに使用することができる．

3.4

risk source

element which alone or in combination has the potential to give rise to *risk* (**3.1**)

3.5

event

occurrence or change of a particular set of circumstances

Note 1 to entry: An event can have one or more occurrences, and can have several causes and several *consequences* (**3.6**).

Note 2 to entry: An event can also be something that is expected which does not happen, or something that is not expected which does happen.

Note 3 to entry: An event can be a risk source.

3.6

consequence

outcome of an *event* (**3.5**) affecting objectives

Note 1 to entry: A consequence can be certain or uncertain and can have positive or negative direct or indirect effects on objectives.

リスク源（risk source）

それ自体又はほかとの組合せによって，リスク（**3.1**）を生じさせる力を潜在的にもっている要素．

3.5

事象（event）

ある一連の周辺状況の出現又は変化．

注記 1　事象は，発生が一度以上であることがあり，幾つかの原因及び幾つかの結果（**3.6**）をもつことがある．

注記 2　事象は，予想していたが起こらないこと，又は予想していなかったが起きることがある．

注記 3　事象がリスク源であることもある．

3.6

結果（consequence）

目的に影響を与える事象（**3.5**）の結末．

注記 1　結果は，確かなことも不確かなこともあり，目的に対して好ましい又は好ましくない直接的影響又は間接的影響を

Note 2 to entry: Consequences can be expressed qualitatively or quantitatively.

Note 3 to entry: Any consequence can escalate through cascading and cumulative effects.

3.7

likelihood

chance of something happening

Note 1 to entry: In *risk management* (**3.2**) terminology, the word "likelihood" is used to refer to the chance of something happening, whether defined, measured or determined objectively or subjectively, qualitatively or quantitatively, and described using general terms or mathematically (such as a probability or a frequency over a given time period).

Note 2 to entry: The English term "likelihood" does not have a direct equivalent in some languages; instead, the equivalent of the term "probability" is

与えることもある.

注記 2　結果は，定性的にも定量的にも表現されることがある.

注記 3　いかなる結果も，波及的影響及び累積的影響によって増大することがある.

3.7

起こりやすさ（likelihood）

何かが起こる可能性.

注記 1　リスクマネジメント（**3.2**）では，“起こりやすさ”という用語は，何かが起こるという可能性を表すために使われる.“起こりやすさ”の定義，測定又は判断は，主観的か若しくは客観的か，又は定性的か若しくは定量的かを問わない.

また，“起こりやすさ”は，一般的な用語を用いて表現するか，又は数学的（例えば，発生確率，所定期間内の頻度など）に表現するかは問わない.

注記 2　幾つかの言語では，英語の“likelihood (起こりやすさ)”と全く同じ意味の語がなく，同義語の“probability（発

often used. However, in English, “probability” is often narrowly interpreted as a mathematical term. Therefore, in risk management terminology, “likelihood” is used with the intent that it should have the same broad interpretation as the term “probability” has in many languages other than English.

3.8
control

measure that maintains and/or modifies *risk* (**3.1**)

Note 1 to entry: Controls include, but are not limited to, any process, policy, device, practice, or other conditions and/or actions which maintain and/or modify risk.

Note 2 to entry: Controls may not always exert the intended or assumed modifying effect.

4 Principles

The purpose of risk management is the creation

生確率)”がしばしば使用される．しかし，英語の“probability”は，数学用語としてしばしば狭義に解釈される．したがって，リスクマネジメント用語では，英語以外の多くの言語において“probability”がもつような幅広い解釈がなされることが望ましいという意図の下で“likelihood”を使用する．

3.8

管理策（control）

リスク（**3.1**）を維持及び／又は修正する対策．

注記 1　管理策は，リスクを維持及び／又は修正するプロセス，方針，方策，実務又はその他の条件及び／若しくは活動を含む．ただし，これらに限定されない．

注記 2　管理策が，常に意図又は想定した修正効果を発揮するとは限らない．

4　原則

リスクマネジメントの意義は，価値の創出及び保

and protection of value. It improves performance, encourages innovation and supports the achievement of objectives.

The principles outlined in **Figure 2** provide guidance on the characteristics of effective and efficient risk management, communicating its value and explaining its intention and purpose. The principles are the foundation for managing risk and should be considered when establishing the organization's risk management framework and processes. These principles should enable an organization to manage the effects of uncertainty on its objectives.

護である．リスクマネジメントは，パフォーマンスを改善し，イノベーションを促進し，目的の達成を支援する．

図 2 に示す原則は，有効かつ効率的なリスクマネジメントの特徴に関する指針を示し，リスクマネジメントの価値を伝え，リスクマネジメントの意図及び意義を説明したものである．原則は，リスクのマネジメントを行うための土台であり，組織のリスクマネジメントの枠組み及びプロセスを確立する際には，原則を検討することが望ましい．不確かさが目的に及ぼす影響のマネジメントを行うことが，これらの原則によって可能になることが望ましい．

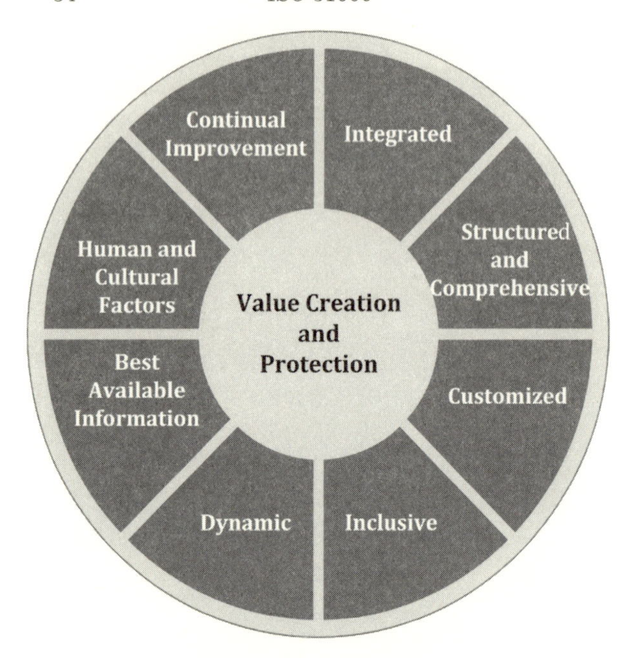

Figure 2 — Principles

Effective risk management requires the elements of **Figure 2** and can be further explained as follows.

a) Integrated

Risk management is an integral part of all organizational activities.

b) Structured and comprehensive

A structured and comprehensive approach to risk management contributes to consistent

図 2 －原則

有効なリスクマネジメントは，**図 2** の要素を要求し，更に次に示すように説明することができる．

a) **統合**　リスクマネジメントは，組織の全ての活動に統合されている．

b) **体系化及び包括**　リスクマネジメントの，体系化され，かつ，包括的な取組み方は，一貫性のある比較可能な結果に寄与する．

and comparable results.

c) Customized

The risk management framework and process are customized and proportionate to the organization's external and internal context related to its objectives.

d) Inclusive

Appropriate and timely involvement of stakeholders enables their knowledge, views and perceptions to be considered. This results in improved awareness and informed risk management.

e) Dynamic

Risks can emerge, change or disappear as an organization's external and internal context changes. Risk management anticipates, detects, acknowledges and responds to those changes and events in an appropriate and timely manner.

f) Best available information

The inputs to risk management are based on historical and current information, as well as on future expectations. Risk management explicitly takes into account any limitations and

c)　**組織への適合**　リスクマネジメントの枠組み及びプロセスは，対象とする組織の，目的に関連する外部及び内部の状況に合わせられ，均衡がとれている．

d)　**包含**　ステークホルダの適切で時宜を得た参画は，彼らの知識，見解及び認識を考慮することを可能にする．これが，意識の向上，及び十分な情報に基づくリスクマネジメントにつながる．

e)　**動的**　組織の外部及び内部の状況の変化に伴って，リスクが出現，変化又は消滅することがある．リスクマネジメントは，これらの変化及び事象を適切に，かつ，時宜を得て予測し，発見し，認識し，それらの変化及び事象に対応する．

f)　**利用可能な最善の情報**　リスクマネジメントへのインプットは，過去及び現在の情報，並びに将来の予想に基づく．リスクマネジメントは，これらの情報及び予想に付随する制約及び不確かさを明確に考慮に入れる．情報は時宜を得て

uncertainties associated with such information and expectations. Information should be timely, clear and available to relevant stakeholders.

g) Human and cultural factors

Human behaviour and culture significantly influence all aspects of risk management at each level and stage.

h) Continual improvement

Risk management is continually improved through learning and experience.

5 Framework

5.1 General

The purpose of the risk management framework is to assist the organization in integrating risk management into significant activities and functions. The effectiveness of risk management will depend on its integration into the governance of the organization, including decision-making. This requires support from stakeholders, particularly top management.

Framework development encompasses integrat-

おり，明確であり，かつ，関連するステークホルダが入手できることが望ましい．

g)　**人的要因及び文化的要因**　人間の行動及び文化は，それぞれのレベル及び段階においてリスクマネジメントの全ての側面に大きな影響を与える．

h)　**継続的改善**　リスクマネジメントは，学習及び経験を通じて継続的に改善される．

5　枠組み

5.1　一般

リスクマネジメントの枠組みの意義は，リスクマネジメントを組織の重要な活動及び機能に統合するときに組織を支援することである．リスクマネジメントの有効性は，意思決定を含む組織統治への統合にかかっている．そのためには，ステークホルダ，特にトップマネジメントの支援が必要である．

枠組みの策定は，組織全体におけるリスクマネジ

ing, designing, implementing, evaluating and improving risk management across the organization. **Figure 3** illustrates the components of a framework.

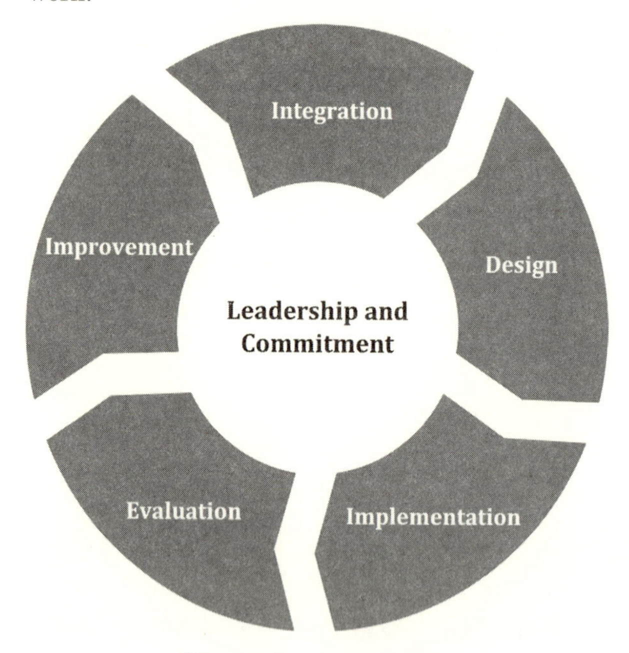

Figure 3 — Framework

The organization should evaluate its existing risk management practices and processes, evaluate any gaps and address those gaps within the framework.

メントの統合，設計，実施，評価及び改善を含む．

図 3 は，枠組みの構成要素を示したものである．

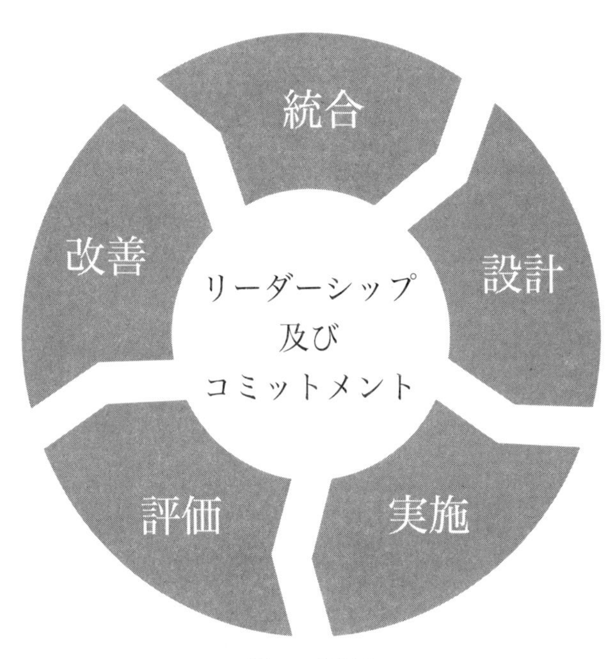

図 3 －枠組み

組織は，既存のリスクマネジメントの方策及びプロセスを評価し，かい離を分析し，枠組みの中でこれらのかい離に取り組むことが望ましい．

The components of the framework and the way in which they work together should be customized to the needs of the organization.

5.2 Leadership and commitment

Top management and oversight bodies, where applicable, should ensure that risk management is integrated into all organizational activities and should demonstrate leadership and commitment by:

— customizing and implementing all components of the framework;

— issuing a statement or policy that establishes a risk management approach, plan or course of action;

— ensuring that the necessary resources are allocated to managing risk;

— assigning authority, responsibility and accountability at appropriate levels within the organization.

This will help the organization to:

— align risk management with its objectives,

枠組みの構成要素と，それらの構成要素が共に機能する仕方は，組織の必要性に合わせて調整することが望ましい．

5.2　リーダーシップ及びコミットメント

トップマネジメント及び監督機関（該当する場合）は，リスクマネジメントが組織の全ての活動に統合されることを確実にすることが望ましい．また，次の事項を通じて，リーダーシップ及びコミットメントを示すことが望ましい．

— 枠組みの全ての要素を組織に合わせて実施する．
— リスクマネジメントの取組み方，計画又は活動方針を確定する声明又は方針を公表する．

— 必要な資源がリスクのマネジメントを行うことに配分されることを確実にする．
— 権限，責任及びアカウンタビリティを，組織内の適切な階層に割り当てる．

リーダーシップ及びコミットメントは，組織の次の事項を促進する．
— リスクマネジメントを，組織の目的，戦略及び

strategy and culture;

— recognize and address all obligations, as well as its voluntary commitments;
— establish the amount and type of risk that may or may not be taken to guide the development of risk criteria, ensuring that they are communicated to the organization and its stakeholders;
— communicate the value of risk management to the organization and its stakeholders;
— promote systematic monitoring of risks;
— ensure that the risk management framework remains appropriate to the context of the organization.

Top management is accountable for managing risk while oversight bodies are accountable for overseeing risk management. Oversight bodies are often expected or required to:

— ensure that risks are adequately considered when setting the organization's objectives;
— understand the risks facing the organization in pursuit of its objectives;
— ensure that systems to manage such risks are

文化と整合させる．

— 組織の全ての義務，及び組織の任意のコミットメントを認識し，これらに取り組む．

— リスク基準の策定の指針として組織が取ることができる，又は取ることができないリスクの大きさ及び種類を確定し，それらのリスクが組織及びステークホルダに伝達されることを確実にする．

— リスクマネジメントの価値を組織及び組織のステークホルダに伝達する．

— リスクの体系的モニタリングを推進する．

— リスクマネジメントの枠組みが組織の状況に対して常に適切であることを確実にする．

トップマネジメントはリスクのマネジメントを行うことに責任を負い，監督機関はリスクマネジメントを監視する責任を負う．監督機関は，しばしば次の事項を行うことを期待され又は必要とされる．

— 組織の目的を決定する際にリスクが十分に検討されることを確実にする．

— 組織が目的の追求に当たって直面するリスクを理解する．

— これらのリスクのマネジメントを行うためのシ

implemented and operating effectively;

— ensure that such risks are appropriate in the context of the organization's objectives;
— ensure that information about such risks and their management is properly communicated.

5.3 Integration

Integrating risk management relies on an understanding of organizational structures and context. Structures differ depending on the organization's purpose, goals and complexity. Risk is managed in every part of the organization's structure. Everyone in an organization has responsibility for managing risk.

Governance guides the course of the organization, its external and internal relationships, and the rules, processes and practices needed to achieve its purpose. Management structures translate governance direction into the strategy and associated objectives required to achieve desired levels of sustainable performance and long-term viability. Determining risk management accountability and

ステムが実施され，有効に運用されることを確実にする．

— 組織の目的に照らして，それらのリスクが適切であることを確実にする．

— それらのリスク及びそれらのマネジメントに関する情報が適切に伝達されることを確実にする．

5.3　統合

リスクマネジメントの統合は，組織の体制及び状況の理解にかかっている．体制は，組織の意図，目標及び複雑さによって異なる．リスクは，組織の体制のあらゆる部分でマネジメントされる．組織の全員が，リスクのマネジメントを行うことに対する責任を負っている．

組織統治は，組織の意図を達成するために，組織の方向性，組織の外部関係及び内部関係，並びに規則，プロセス及び方策を導く．経営体制は，組織統治の方向性を，望ましいレベルの持続可能なパフォーマンス及び長期的な継続性を達成するために必要な戦略及び関連する目的へと転換する．組織内部におけるアカウンタビリティ及び監視の役割の決定は，組織の統治の不可欠な部分である．

oversight roles within an organization are integral parts of the organization's governance.

Integrating risk management into an organization is a dynamic and iterative process, and should be customized to the organization's needs and culture. Risk management should be a part of, and not separate from, the organizational purpose, governance, leadership and commitment, strategy, objectives and operations.

5.4 Design

5.4.1 Understanding the organization and its context

When designing the framework for managing risk, the organization should examine and understand its external and internal context.

Examining the organization's external context may include, but is not limited to:

— the social, cultural, political, legal, regulatory, financial, technological, economic and environmental factors, whether international, national, regional or local;

リスクマネジメントと組織との統合は，動的かつ反復的なプロセスである．この統合は，組織の必要性及び文化に合わせることが望ましい．リスクマネジメントは，組織の意図，組織統治，リーダーシップ及びコミットメント，戦略，目的並びに業務活動の一部となり，これらと分離していないことが望ましい．

5.4　設計

5.4.1　組織及び組織の状況の理解

リスクのマネジメントを行うための枠組みを設計するに当たって，組織は，外部及び内部の状況を検証し，理解することが望ましい．

組織の外部状況の検証には，次の事項が含まれる場合がある．ただし，これらに限らない．

— 国際，国内，地方又は近隣地域を問わず，社会，文化，政治，法律，規制，金融，技術，経済及び環境に関する要因

— key drivers and trends affecting the objectives of the organization;
— external stakeholders' relationships, perceptions, values, needs and expectations;
— contractual relationships and commitments;
— the complexity of networks and dependencies.

Examining the organization's internal context may include, but is not limited to:

— vision, mission and values;
— governance, organizational structure, roles and accountabilities;
— strategy, objectives and policies;
— the organization's culture;
— standards, guidelines and models adopted by the organization;
— capabilities, understood in terms of resources and knowledge (e.g. capital, time, people, intellectual property, processes, systems and technologies);
— data, information systems and information flows;
— relationships with internal stakeholders, taking into account their perceptions and values;

— 組織の目的に影響を与える，鍵となる原動力及び傾向
— 外部ステークホルダとの関係，並びに外部ステークホルダの認知，価値観，必要性及び期待
— 契約上の関係及びコミットメント
— ネットワークの複雑さ，及び依存関係

組織の内部状況の検証には，次の事項が含まれる場合がある．ただし，これらに限らない．

— ビジョン，使命及び価値観
— 組織統治，組織体制，役割及びアカウンタビリティ
— 戦略，目的及び方針
— 組織の文化
— 組織が採用する規格，指針及びモデル
— 資源及び知識として理解される能力（例えば，資本，時間，人員，知的財産，プロセス，システム，技術）
— データ，情報システム及び情報の流れ
— 内部ステークホルダの認知及び価値観を考慮に入れた，内部ステークホルダとの関係

— contractual relationships and commitments;
— interdependencies and interconnections.

5.4.2 Articulating risk management commitment

Top management and oversight bodies, where applicable, should demonstrate and articulate their continual commitment to risk management through a policy, a statement or other forms that clearly convey an organization's objectives and commitment to risk management. The commitment should include, but is not limited to:

— the organization's purpose for managing risk and links to its objectives and other policies;
— reinforcing the need to integrate risk management into the overall culture of the organization;
— leading the integration of risk management into core business activities and decision-making;
— authorities, responsibilities and accountabilities;
— making the necessary resources available;

— 契約上の関係及びコミットメント
— 相互依存及び相互関連

5.4.2 リスクマネジメントに関するコミットメントの明示

トップマネジメント及び監督機関（該当する場合）は，リスクマネジメントに対する継続的なコミットメントを行動で示し，明示することが望ましい．これは，組織の目的及びリスクマネジメントへのコミットメントを明確に伝える方針，声明又はその他の形式で行うことができる．コミットメントには，次の事項を含めることが望ましい．ただし，これらに限らない．

— 組織がリスクのマネジメントを行う意義，並びに組織の目的及びその他の方針とのつながり
— リスクマネジメントを組織全体の文化に統合する必要性を強めること
— リスクマネジメントと中核的事業活動及び意思決定との統合を主導すること
— 権限，責任及びアカウンタビリティ
— 必要な資源を利用可能にすること

— the way in which conflicting objectives are dealt with;
— measurement and reporting within the organization's performance indicators;
— review and improvement.

The risk management commitment should be communicated within an organization and to stakeholders, as appropriate.

5.4.3 Assigning organizational roles, authorities, responsibilities and accountabilities

Top management and oversight bodies, where applicable, should ensure that the authorities, responsibilities and accountabilities for relevant roles with respect to risk management are assigned and communicated at all levels of the organization, and should:

— emphasize that risk management is a core responsibility;
— identify individuals who have the accountability and authority to manage risk (risk owners).

— 相反する目的への対処の仕方

— 組織のパフォーマンス指標の中での測定及び報告
— レビュー及び改善

リスクマネジメントに関するコミットメントを，必要に応じて，組織内及びステークホルダに伝達することが望ましい．

5.4.3　組織の役割，権限，責任及びアカウンタビリティの割当て

トップマネジメント及び監督機関（該当する場合）は，リスクマネジメントに関して，関連する役割のアカウンタビリティ，責任及び権限が組織のあらゆる階層で割り当てられ，伝達されることを確実にし，次の事項を行うことが望ましい．

— リスクマネジメントは，中核的な責務であることを強調する．
— リスクのマネジメントを行うためのアカウンタビリティ及び権限をもつ個人（リスク所有者）を特定する．

5.4.4 Allocating resources

Top management and oversight bodies, where applicable, should ensure allocation of appropriate resources for risk management, which can include, but are not limited to:

— people, skills, experience and competence;
— the organization's processes, methods and tools to be used for managing risk;
— documented processes and procedures;
— information and knowledge management systems;
— professional development and training needs.

The organization should consider the capabilities of, and constraints on, existing resources.

5.4.5 Establishing communication and consultation

The organization should establish an approved approach to communication and consultation in order to support the framework and facilitate the effective application of risk management. Communication involves sharing information with targeted

5.4.4　資源の配分

トップマネジメント及び監督機関（該当する場合）は，リスクマネジメントのための適切な資源の割当てを確実にすることが望ましい．資源には，次の事項が含まれる場合がある．ただし，これらに限らない．

— 人員，技能，経験及び力量
— リスクのマネジメントを行うために使用する，組織のプロセス，方法及び手段
— 文書化されたプロセス及び手順
— 情報及び知識のマネジメントシステム
— 専門的な人材開発及び教育訓練の必要性

組織は，既存の資源の能力及び制約要因を考慮することが望ましい．

5.4.5　コミュニケーション及び協議の確立

組織は，枠組みを支え，リスクマネジメントの効果的な適用を促進するために，コミュニケーション及び協議に対する，認められた取組み方を確立することが望ましい．コミュニケーションは，対象者とする相手との情報共有を含む．また，協議は，意思

audiences. Consultation also involves participants providing feedback with the expectation that it will contribute to and shape decisions or other activities. Communication and consultation methods and content should reflect the expectations of stakeholders, where relevant.

Communication and consultation should be timely and ensure that relevant information is collected, collated, synthesised and shared, as appropriate, and that feedback is provided and improvements are made.

5.5 Implementation

The organization should implement the risk management framework by:

— developing an appropriate plan including time and resources;
— identifying where, when and how different types of decisions are made across the organization, and by whom;
— modifying the applicable decision-making processes where necessary;
— ensuring that the organization's arrange-

決定又はその他の活動に寄与し，これらを形成することを期待してフィードバックを提供する参加者をも含む．関連する場合，コミュニケーション及び協議の方法及び内容は，ステークホルダの期待を反映することが望ましい．

コミュニケーション及び協議は，適時に行うことが望ましい．また，関連する情報が適切に収集され，照合され，統合され，共有されること，及びフィードバックが提供され，改善がなされることを確実にすることが望ましい．

5.5　実施

組織は，次の事項を行うことによって，リスクマネジメントの枠組みを実施することが望ましい．

— 時間及び資源を含めた適切な計画を策定する．

— 様々な種類の決定が，組織全体のどこで，いつ，どのように，また，誰によって下されるのかを特定する．
— 必要に応じて，適用される意思決定プロセスを修正する．
— リスクのマネジメントを行うことに関する組織

ments for managing risk are clearly understood and practised.

Successful implementation of the framework requires the engagement and awareness of stakeholders. This enables organizations to explicitly address uncertainty in decision-making, while also ensuring that any new or subsequent uncertainty can be taken into account as it arises.

Properly designed and implemented, the risk management framework will ensure that the risk management process is a part of all activities throughout the organization, including decision-making, and that changes in external and internal contexts will be adequately captured.

5.6 Evaluation

In order to evaluate the effectiveness of the risk management framework, the organization should:

— periodically measure risk management framework performance against its purpose, implementation plans, indicators and expected behaviour;

の取決めが明確に理解され，実施されることを確実にする．

枠組みの実施を成功させるためには，ステークホルダが参画し，自ら認識することが必要である．これによって，組織は，新たな不確かさ又は後続の不確かさが発生する都度，それらを考慮に入れることを可能にし，また，意思決定において不確かさに明確な形で取り組むことができる．

適切に設計され，実施されたリスクマネジメントの枠組みは，リスクマネジメントプロセスが，意思決定を含め，組織全体の全ての活動の一部になること，並びに外部及び内部の状況の変化が適切に取り入れられることを確実にする．

5.6　評価

リスクマネジメントの枠組みの有効性を評価するために，組織は，次の事項を行うことが望ましい．

— 意義，実施計画，指標及び期待される行動に照らして，リスクマネジメントの枠組みのパフォーマンスを定期的に測定する．

— determine whether it remains suitable to support achieving the objectives of the organization.

5.7 Improvement

5.7.1 Adapting

The organization should continually monitor and adapt the risk management framework to address external and internal changes. In doing so, the organization can improve its value.

5.7.2 Continually improving

The organization should continually improve the suitability, adequacy and effectiveness of the risk management framework and the way the risk management process is integrated.

As relevant gaps or improvement opportunities are identified, the organization should develop plans and tasks and assign them to those accountable for implementation. Once implemented, these improvements should contribute to the enhancement of risk management.

— リスクマネジメントの枠組みが組織の目的達成を支援するために適した状態か否かを明確にする.

5.7　改善

5.7.1　適応

組織は，外部及び内部の変化に対応できるように，リスクマネジメントの枠組みを継続的にモニタリングし，適応させることが望ましい．それによって，組織は自らの価値を高めることができる.

5.7.2　継続的改善

組織は，リスクマネジメントの枠組みの適切性，妥当性及び有効性，並びにリスクマネジメントプロセスを統合する方法を継続的に改善することが望ましい.

関連するかい離又は改善の機会が特定された時点で，組織は計画及び実施事項を策定し，実施に関してアカウンタビリティをもつ人にそれらを割り当てることが望ましい．これらの改善は，実施された時点でリスクマネジメントの向上に寄与するはずである.

6 Process

6.1 General

The risk management process involves the systematic application of policies, procedures and practices to the activities of communicating and consulting, establishing the context and assessing, treating, monitoring, reviewing, recording and reporting risk. This process is illustrated in **Figure 4**.

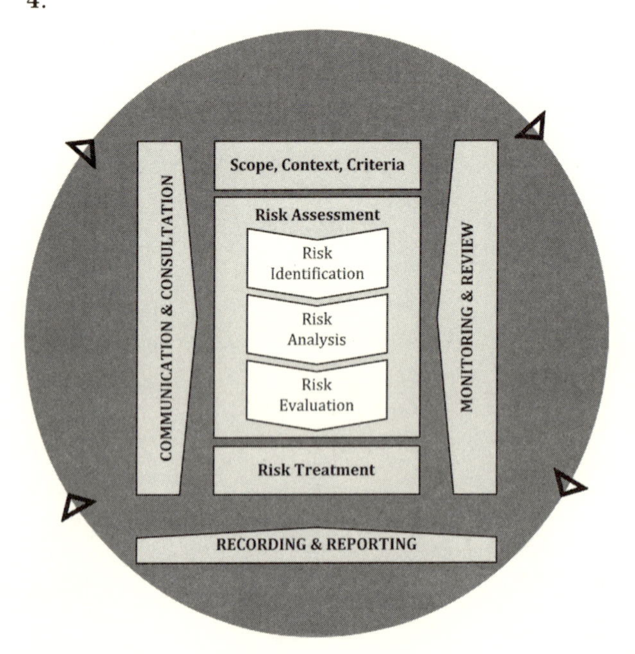

Figure 4 — Process

6 プロセス

6.1 一般

リスクマネジメントプロセスには，方針，手順及び方策を，コミュニケーション及び協議，状況の確定，並びにリスクのアセスメント，対応，モニタリング，レビュー，記録作成及び報告の活動に体系的に適用することが含まれる．このプロセスを**図 4**に示す．

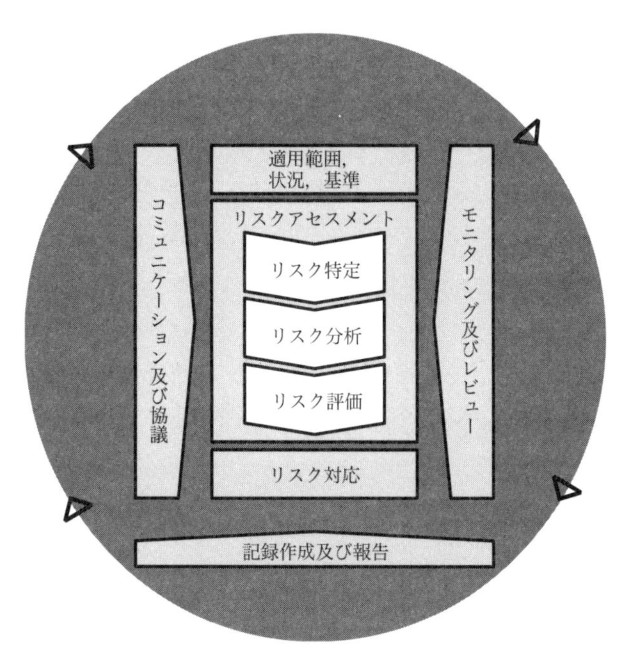

図 4 －プロセス

The risk management process should be an integral part of management and decision-making and integrated into the structure, operations and processes of the organization. It can be applied at strategic, operational, programme or project levels.

There can be many applications of the risk management process within an organization, customized to achieve objectives and to suit the external and internal context in which they are applied.

The dynamic and variable nature of human behaviour and culture should be considered throughout the risk management process.

Although the risk management process is often presented as sequential, in practice it is iterative.

6.2 Communication and consultation

The purpose of communication and consultation is to assist relevant stakeholders in understanding risk, the basis on which decisions are made and the reasons why particular actions are required. Com-

リスクマネジメントプロセスは，マネジメント及び意思決定における不可欠な部分であることが望ましい．また，組織の体制，業務活動及びプロセスに組み込まれていることが望ましい．リスクマネジメントプロセスは，戦略，業務活動，プログラム又はプロジェクトの段階で適用することができる．

組織の目的を達成することに合わせ，かつ，適用される外部及び内部の状況に適応するために，組織の中で，リスクマネジメントプロセスが，多数適用されている場合がある．

リスクマネジメントプロセス全体にわたって，人間の行動及び文化がもつ動的で可変的な性質を考慮することが望ましい．

リスクマネジメントプロセスは，しばしば逐次的なものとして表されるが，実務では反復的である．

6.2 コミュニケーション及び協議

コミュニケーション及び協議の意義は，関連するステークホルダが，リスク，意思決定の根拠，及び特定の活動が必要な理由が理解できるように支援することである．コミュニケーションは，リスクに対

munication seeks to promote awareness and understanding of risk, whereas consultation involves obtaining feedback and information to support decision-making. Close coordination between the two should facilitate factual, timely, relevant, accurate and understandable exchange of information, taking into account the confidentiality and integrity of information as well as the privacy rights of individuals.

Communication and consultation with appropriate external and internal stakeholders should take place within and throughout all steps of the risk management process.

Communication and consultation aims to:

— bring different areas of expertise together for each step of the risk management process;
— ensure that different views are appropriately considered when defining risk criteria and when evaluating risks;
— provide sufficient information to facilitate risk oversight and decision-making;

する意識及び理解の促進を目指す．一方，協議は，意思決定を裏付けるためのフィードバック及び情報の入手を含む．コミュニケーションと協議とを密接に組み合わせることによって，情報の機密性及び完全性，並びに個人のプライバシー権を考慮しながら，事実に基づく，時宜を得た，適切で正確かつ理解可能な情報交換が促進される．

適切な外部及び内部のステークホルダとのコミュニケーション及び協議は，リスクマネジメントプロセスの各段階及び全体で実施することが望ましい．

コミュニケーション及び協議の狙いは，次のとおりである．

— リスクマネジメントプロセスの各段階に関して，異なった領域の専門知識を集める．
— リスク基準を定め，リスクを評価する場合には，異なった見解について適切に考慮することを確実にする．
— リスク監視及び意思決定を促進するために十分な情報を提供する．

— build a sense of inclusiveness and ownership among those affected by risk.

6.3 Scope, context and criteria

6.3.1 General

The purpose of establishing the scope, the context and criteria is to customize the risk management process, enabling effective risk assessment and appropriate risk treatment. Scope, context and criteria involve defining the scope of the process, and understanding the external and internal context.

6.3.2 Defining the scope

The organization should define the scope of its risk management activities.

As the risk management process may be applied at different levels (e.g. strategic, operational, programme, project, or other activities), it is important to be clear about the scope under consideration, the relevant objectives to be considered and their alignment with organizational objectives.

When planning the approach, considerations in-

— リスクの影響を受ける者たちの間に一体感及び当事者意識を構築する.

6.3 適用範囲，状況及び基準

6.3.1 一般

適用範囲，状況及び基準を確定する意義は，リスクマネジメントプロセスを組織に合わせ，効果的なリスクアセスメント及び適切なリスク対応を可能にすることである．適用範囲，状況及び基準は，プロセスの適用範囲を定め，外部及び内部の状況を理解することを含む.

6.3.2 適用範囲の決定

組織は，リスクマネジメント活動の適用範囲を定めることが望ましい.

リスクマネジメントプロセスは，様々なレベル（例えば，戦略，業務活動，プログラム，プロジェクト又はその他の活動）で適用されるため，検討の対象となる適用範囲，検討の対象となる関連目的，並びにそれらと組織の目的との整合を明確にすることが重要である.

取組み方を計画する際の検討事項は，次を含む.

clude:

— objectives and decisions that need to be made;
— outcomes expected from the steps to be taken in the process;
— time, location, specific inclusions and exclusions;
— appropriate risk assessment tools and techniques;
— resources required, responsibilities and records to be kept;
— relationships with other projects, processes and activities.

6.3.3 External and internal context

The external and internal context is the environment in which the organization seeks to define and achieve its objectives.

The context of the risk management process should be established from the understanding of the external and internal environment in which the organization operates and should reflect the specific environment of the activity to which the risk management process is to be applied.

— 目的，及び下す必要のある決定
— プロセスにおいてとられる対策によって期待される結末
— 時間，場所，個々の包含及び除外

— 適切なリスクアセスメントの手段及び手法

— 必要とされる資源，責任，及び残すべき記録

— 他のプロジェクト，プロセス及び活動との関係

6.3.3 外部及び内部の状況

外部及び内部の状況とは，組織が自らの目的を定め，その目的を達成しようとする状態を取り巻く環境である．

リスクマネジメントプロセスの状況は，組織が業務活動を行う外部及び内部の環境の理解から確定されることが望ましい．また，リスクマネジメントプロセスが適用される活動の個々の環境を反映することが望ましい．

Understanding the context is important because:

— risk management takes place in the context of the objectives and activities of the organization;

— organizational factors can be a source of risk;

— the purpose and scope of the risk management process may be interrelated with the objectives of the organization as a whole.

The organization should establish the external and internal context of the risk management process by considering the factors mentioned in **5.4.1**.

6.3.4 Defining risk criteria

The organization should specify the amount and type of risk that it may or may not take, relative to objectives. It should also define criteria to evaluate the significance of risk and to support decision-making processes. Risk criteria should be aligned with the risk management framework and customized to the specific purpose and scope of the activity under consideration. Risk criteria should reflect the organization's values, objectives and resources and be consistent with policies and state-

状況の理解は，次に示す理由で重要である．

— リスクマネジメントは，組織の目的及び活動に沿って実施される．

— 組織要因がリスク源になることがある．

— リスクマネジメントプロセスの意義及び範囲が，組織全体の目的と相互に関係していることがある．

— 組織は，**5.4.1** に挙げた要因を考慮することによって，リスクマネジメントプロセスの外部及び内部の状況を確立することが望ましい．

6.3.4　リスク基準の決定

組織は，目的に照らして，取ってもよいリスク又は取ってはならないリスクの大きさ及び種類を規定することが望ましい．組織はまた，リスクの重大性を評価し，意思決定プロセスを支援するための基準を決定することが望ましい．リスク基準は，リスクマネジメントの枠組みと整合させ，検討対象になっている活動に特有の意義及び範囲にリスク基準を合わせることが望ましい．リスク基準は，組織の価値観，目的及び資源を反映し，リスクマネジメント方針及び声明と一致していることが望ましい．基準

ments about risk management. The criteria should be defined taking into consideration the organization's obligations and the views of stakeholders.

While risk criteria should be established at the beginning of the risk assessment process, they are dynamic and should be continually reviewed and amended, if necessary.

To set risk criteria, the following should be considered:

— the nature and type of uncertainties that can affect outcomes and objectives (both tangible and intangible);
— how consequences (both positive and negative) and likelihood will be defined and measured;
— time-related factors;
— consistency in the use of measurements;
— how the level of risk is to be determined;
— how combinations and sequences of multiple risks will be taken into account;
— the organization's capacity.

は，組織の義務及びステークホルダの見解を考慮に入れて規定することが望ましい．

リスク基準は，リスクアセスメントプロセスの開始時に確定することが望ましいが，リスク基準は動的であるため，継続的にレビューを行い，必要に応じて修正することが望ましい．

リスク基準を設定するに当たっては，次の事項を考慮することが望ましい．

— 結末及び目的（有形及び無形の両方）に影響を与える不確かさの特質及び種類

— 結果（好ましい結果及び好ましくない結果の両方）及び起こりやすさをどのように定め，また，測定するか．
— 時間に関連する要素
— 測定法の一貫性
— リスクレベルをどのように決定するか．
— 複数のリスクの組合せ及び順序をどのように考慮に入れるか．
— 組織の能力

6.4 Risk assessment

6.4.1 General

Risk assessment is the overall process of risk identification, risk analysis and risk evaluation.

Risk assessment should be conducted systematically, iteratively and collaboratively, drawing on the knowledge and views of stakeholders. It should use the best available information, supplemented by further enquiry as necessary.

6.4.2 Risk identification

The purpose of risk identification is to find, recognize and describe risks that might help or prevent an organization achieving its objectives. Relevant, appropriate and up-to-date information is important in identifying risks.

The organization can use a range of techniques for identifying uncertainties that may affect one or more objectives. The following factors, and the relationship between these factors, should be considered:

— tangible and intangible sources of risk;

6.4 リスクアセスメント

6.4.1 一般

リスクアセスメントとは，リスク特定，リスク分析及びリスク評価を網羅するプロセス全体を指す．

リスクアセスメントは，ステークホルダの知識及び見解を生かし，体系的，反復的，協力的に行われることが望ましい．必要に応じて，追加的な調査で補完し，利用可能な最善の情報を使用することが望ましい．

6.4.2 リスク特定

リスク特定の意義は，組織の目的の達成を助ける又は妨害する可能性のあるリスクを発見し，認識し，記述することである．リスクの特定に当たっては，現況に即した，適切で最新の情報が重要である．

組織は，一つ以上の目的に影響するかもしれない不確かさを特定するために，様々な手法を使用することができる．次の要素，及びこれらの要素間の関係を考慮することが望ましい．

— 有形及び無形のリスク源

- causes and events;
- threats and opportunities;
- vulnerabilities and capabilities;
- changes in the external and internal context;
- indicators of emerging risks;
- the nature and value of assets and resources;
- consequences and their impact on objectives;
- limitations of knowledge and reliability of information;
- time-related factors;
- biases, assumptions and beliefs of those involved.

The organization should identify risks, whether or not their sources are under its control. Consideration should be given that there may be more than one type of outcome, which may result in a variety of tangible or intangible consequences.

6.4.3 Risk analysis

The purpose of risk analysis is to comprehend the nature of risk and its characteristics including, where appropriate, the level of risk. Risk analysis involves a detailed consideration of uncertainties,

— 原因及び事象
— 脅威及び機会
— ぜい（脆）弱性及び能力
— 外部及び内部の状況の変化
— 新たに発生するリスクの指標
— 資産及び組織の資源の性質及び価値
— 結果及び結果が目的に与える影響
— 知識の限界及び情報の信頼性

— 時間に関連する要素
— 関与する人の先入観，前提及び信条

組織は，リスク源が組織の管理下にあるか否かを問わず，リスクを特定することが望ましい．様々な有形又は無形の結果をもたらす可能性のある2種類以上の結末が存在するかもしれないことを考慮することが望ましい．

6.4.3　リスク分析

リスク分析の意義は，必要に応じてリスクのレベルを含め，リスクの性質及び特徴を理解することである．リスク分析には，不確かさ，リスク源，結果，起こりやすさ，事象，シナリオ，管理策及び管

risk sources, consequences, likelihood, events, scenarios, controls and their effectiveness. An event can have multiple causes and consequences and can affect multiple objectives.

Risk analysis can be undertaken with varying degrees of detail and complexity, depending on the purpose of the analysis, the availability and reliability of information, and the resources available. Analysis techniques can be qualitative, quantitative or a combination of these, depending on the circumstances and intended use.

Risk analysis should consider factors such as:

— the likelihood of events and consequences;
— the nature and magnitude of consequences;
— complexity and connectivity;
— time-related factors and volatility;
— the effectiveness of existing controls;
— sensitivity and confidence levels.

The risk analysis may be influenced by any divergence of opinions, biases, perceptions of risk and

理策の有効性の詳細な検討が含まれる．一つの事象が複数の原因及び結果をもち，複数の目的に影響を与えることがある．

リスク分析は，分析の意義，情報の入手可能性及び信頼性，並びに利用可能な資源に応じて，様々な詳細さ及び複雑さの度合いで行うことができる．分析手法は，周辺状況及び意図する用途に応じて，定性的，定量的，又はそれらを組み合わせたものにすることができる．

リスク分析では，例えば，次の要素を検討することが望ましい．

— 事象の起こりやすさ及び結果
— 結果の性質及び大きさ
— 複雑さ及び結合性
— 時間に関係する要素及び変動性
— 既存の管理策の有効性
— 機微性及び機密レベル

リスク分析は，意見の相違，先入観，リスクの認知及び判断によって影響されることがある．その他

judgements. Additional influences are the quality of the information used, the assumptions and exclusions made, any limitations of the techniques and how they are executed. These influences should be considered, documented and communicated to decision makers.

Highly uncertain events can be difficult to quantify. This can be an issue when analysing events with severe consequences. In such cases, using a combination of techniques generally provides greater insight.

Risk analysis provides an input to risk evaluation, to decisions on whether risk needs to be treated and how, and on the most appropriate risk treatment strategy and methods. The results provide insight for decisions, where choices are being made, and the options involve different types and levels of risk.

6.4.4 Risk evaluation

The purpose of risk evaluation is to support decisions. Risk evaluation involves comparing the re-

の影響としては，使用する情報の質，加えられた前提及び除外された前提，手法の限界，並びに実行方法が挙げられる．これらの影響を検討し，文書化し，意思決定者に伝達することが望ましい．

非常に不確かな事象は，定量化が困難なことがある．重大な結果をもたらす事象を分析する場合，これは課題になる．このような場合は，一般的に手法の組合せを用いることによって洞察が深まる．

リスク分析は，リスク評価へのインプット，リスク対応の必要性及び方法，並びに最適なリスク対応の戦略及び方法の決定へのインプットを提供する．結果は，選択を行う場合に決定を下すための洞察力を提供する．また，選択肢は，様々な種類及びレベルのリスクを伴う．

6.4.4　リスク評価

リスク評価の意義は，決定を裏付けることである．リスク評価は，どこに追加の行為をとるかを決

sults of the risk analysis with the established risk criteria to determine where additional action is required. This can lead to a decision to:

— do nothing further;
— consider risk treatment options;
— undertake further analysis to better understand the risk;
— maintain existing controls;
— reconsider objectives.

Decisions should take account of the wider context and the actual and perceived consequences to external and internal stakeholders.

The outcome of risk evaluation should be recorded, communicated and then validated at appropriate levels of the organization.

6.5 Risk treatment

6.5.1 General

The purpose of risk treatment is to select and implement options for addressing risk.

Risk treatment involves an iterative process of:

定するために，リスク分析の結果と確立されたリスク基準との比較を含む．これによって，次の事項の決定がもたらされる．

— 更なる活動は行わない．
— リスク対応の選択肢を検討する．
— リスクをより深く理解するために，更なる分析に着手する．
— 既存の管理策を維持する．
— 目的を再考する．

意思決定では，より広い範囲の状況，並びに外部及び内部のステークホルダにとっての実際の結果及び認知された結果を考慮することが望ましい．

組織の適切なレベルで，リスク評価の結果を記録し，伝達し，更に検証することが望ましい．

6.5　リスク対応

6.5.1　一般

リスク対応の意義は，リスクに対処するための選択肢を選定し，実施することである．

リスク対応には，次の事項の反復的プロセスが含

— formulating and selecting risk treatment options;
— planning and implementing risk treatment;
— assessing the effectiveness of that treatment;
— deciding whether the remaining risk is acceptable;
— if not acceptable, taking further treatment.

6.5.2 Selection of risk treatment options

Selecting the most appropriate risk treatment option(s) involves balancing the potential benefits derived in relation to the achievement of the objectives against costs, effort or disadvantages of implementation.

Risk treatment options are not necessarily mutually exclusive or appropriate in all circumstances. Options for treating risk may involve one or more of the following:

— avoiding the risk by deciding not to start or continue with the activity that gives rise to the risk;
— taking or increasing the risk in order to pur-

まれる.

— リスク対応の選択肢の策定及び選定

— リスク対応の計画及び実施

— その対応の有効性の評価

— 残留リスクが許容可能かどうかの判断

— 許容できない場合は，更なる対応の実施

6.5.2 リスク対応の選択肢の選定

最適なリスク対応の選択肢の選定には，目的の達成に関して得られる便益と，実施の費用，労力又は不利益との均衡をとることが含まれる.

リスク対応の選択肢は，必ずしも相互に排他的なものではなく，また，全ての周辺状況に適切であるとは限らない．リスク対応の選択肢には，次の事項の一つ以上が含まれてもよい.

— リスクを生じさせる活動を開始又は継続しないと決定することによってリスクを回避する.

— ある機会を追求するために，リスクを取る又は

sue an opportunity;
- removing the risk source;
- changing the likelihood;
- changing the consequences;
- sharing the risk (e.g. through contracts, buying insurance);
- retaining the risk by informed decision.

Justification for risk treatment is broader than solely economic considerations and should take into account all of the organization's obligations, voluntary commitments and stakeholder views. The selection of risk treatment options should be made in accordance with the organization's objectives, risk criteria and available resources.

When selecting risk treatment options, the organization should consider the values, perceptions and potential involvement of stakeholders and the most appropriate ways to communicate and consult with them. Though equally effective, some risk treatments can be more acceptable to some stakeholders than to others.

増加させる.

— リスク源を除去する.
— 起こりやすさを変える.
— 結果を変える.
— (例えば, 契約, 保険購入によって) リスクを共有する.
— 情報に基づいた意思決定によって, リスクを保有する.

リスク対応の根拠は, 単なる経済的な考慮事項より幅広いため, 組織の義務, 任意のコミットメント及びステークホルダの見解の全てを考慮に入れることが望ましい. リスク対応の選択肢の選定は, 組織の目的, リスク基準及び利用可能な資源に基づいて行われることが望ましい.

リスク対応の選択肢を選定する際に, 組織は, ステークホルダの価値観, 認知及び関与の可能性, 並びにステークホルダとのコミュニケーション及び協議に最適な仕方を考慮することが望ましい. 有効性は同じでも, ステークホルダによってリスク対応策の受け入れやすさは異なることがある.

Risk treatments, even if carefully designed and implemented might not produce the expected outcomes and could produce unintended consequences. Monitoring and review need to be an integral part of the risk treatment implementation to give assurance that the different forms of treatment become and remain effective.

Risk treatment can also introduce new risks that need to be managed.

If there are no treatment options available or if treatment options do not sufficiently modify the risk, the risk should be recorded and kept under ongoing review.

Decision makers and other stakeholders should be aware of the nature and extent of the remaining risk after risk treatment. The remaining risk should be documented and subjected to monitoring, review and, where appropriate, further treatment.

6.5.3 Preparing and implementing risk

慎重に設計し，実施したとしても，リスク対応は予想した結末を生まないかもしれないし，意図しない結果をもたらすこともある．様々な形態のリスク対応を有効にし，その有効性が維持されることを保証するためには，モニタリング及びレビューをリスク対応実施の一体部分とする必要がある．

リスク対応が，新たにマネジメントを行うことが必要なリスクをもたらす可能性もある．

利用可能なリスク対応の選択肢がない場合，又はリスク対応の選択肢によってリスクが十分に変化しない場合には，そのリスクを記録し，継続的なレビューの対象とすることが望ましい．

意思決定者及びその他のステークホルダは，リスク対応後の残留リスクの性質及び程度を知ることが望ましい．残留リスクは，文書化し，モニタリングし及びレビューし，並びに必要に応じて追加的対応の対象とすることが望ましい．

6.5.3　リスク対応計画の準備及び実施

treatment plans

The purpose of risk treatment plans is to specify how the chosen treatment options will be implemented, so that arrangements are understood by those involved, and progress against the plan can be monitored. The treatment plan should clearly identify the order in which risk treatment should be implemented.

Treatment plans should be integrated into the management plans and processes of the organization, in consultation with appropriate stakeholders.

The information provided in the treatment plan should include:

— the rationale for selection of the treatment options, including the expected benefits to be gained;
— those who are accountable and responsible for approving and implementing the plan;
— the proposed actions;
— the resources required, including contingencies;

リスク対応計画の意義は，関与する人々が取決めを理解し，計画に照らして進捗状況をモニタリングできるように，選定した対応選択肢をどのように実施するかを規定することである．対応計画には，リスク対応を実施する順序を明記することが望ましい．

対応計画は，適切なステークホルダと協議の上で，組織の経営計画及びプロセスに統合されることが望ましい．

対応計画で提供される情報には，次の事項を含めることが望ましい．

— 期待される取得便益を含めた，対応選択肢の選定の理由

— 計画の承認及び実施に関してアカウンタビリティ及び責任をもつ人

— 提案された活動

— 不測の事態への対応を含む，必要とされる資源

— the performance measures;
— the constraints;
— the required reporting and monitoring;
— when actions are expected to be undertaken and completed.

6.6 Monitoring and review

The purpose of monitoring and review is to assure and improve the quality and effectiveness of process design, implementation and outcomes. Ongoing monitoring and periodic review of the risk management process and its outcomes should be a planned part of the risk management process, with responsibilities clearly defined.

Monitoring and review should take place in all stages of the process. Monitoring and review includes planning, gathering and analysing information, recording results and providing feedback.

The results of monitoring and review should be incorporated throughout the organization's performance management, measurement and reporting activities.

— パフォーマンスの尺度
— 制約要因
— 必要な報告及びモニタリング
— 活動が実行され，完了することが予想される時期

6.6 モニタリング及びレビュー

モニタリング及びレビューの意義は，プロセスの設計，実施及び結末の質及び効果を保証し，改善することである．責任を明確に定めた上で，リスクマネジメントプロセス及びその結末の継続的モニタリング及び定期的レビューを，リスクマネジメントプロセスの計画的な部分とすることが望ましい．

モニタリング及びレビューは，プロセスの全ての段階で行うことが望ましい．モニタリング及びレビューは，計画，情報の収集及び分析，結果の記録作成，並びにフィードバックの提供を含む．

モニタリング及びレビューの結果が，組織のパフォーマンスマネジメント，測定及び報告活動全体に組み込まれることが望ましい．

6.7 Recording and reporting

The risk management process and its outcomes should be documented and reported through appropriate mechanisms. Recording and reporting aims to:

— communicate risk management activities and outcomes across the organization;
— provide information for decision-making;
— improve risk management activities;
— assist interaction with stakeholders, including those with responsibility and accountability for risk management activities.

Decisions concerning the creation, retention and handling of documented information should take into account, but not be limited to: their use, information sensitivity and the external and internal context.

Reporting is an integral part of the organization's governance and should enhance the quality of dialogue with stakeholders and support top management and oversight bodies in meeting their responsibilities. Factors to consider for reporting include,

6.7　記録作成及び報告

適切な仕組みを通じて，リスクマネジメントプロセス及びその結末を文書化し，報告することが望ましい．記録作成及び報告の狙いは，次のとおりである．

— 組織全体にリスクマネジメント活動及び結末を伝達する．
— 意思決定のための情報を提供する．
— リスクマネジメント活動を改善する．
— リスクマネジメント活動の責任及びアカウンタビリティをもつ人々を含めたステークホルダとのやり取りを補助する．

文書化した情報の作成，保持及び取扱いに関する意思決定に際しては，情報の用途，情報の機微性，並びに外部及び内部の状況を考慮することが望ましいが，考慮する事項はこれらに限らない．

報告は，組織の統治の不可欠な部分であり，ステークホルダとの会話の質を高め，トップマネジメント及び監督機関が責任を果たすことができるように支援することが望ましい．報告に当たって考慮すべき要素には，次の事項が含まれる．ただし，これら

but are not limited to:

— differing stakeholders and their specific information needs and requirements;
— cost, frequency and timeliness of reporting;
— method of reporting;
— relevance of information to organizational objectives and decision-making.

に限らない．

— 様々なステークホルダ，並びにそれらのステークホルダに特有の情報の必要性及び要求事項
— 報告の費用，頻度及び適時性
— 報告の方法
— 情報と組織の目的及び意思決定との関連性

Bibliography

[1] IEC 31010, *Risk management — Risk assessment techniques*

参考文献

[1] **JIS Q 0073**:2010　リスクマネジメント－用語

[2] **JIS Q 31010**:2012　リスクマネジメント－リスクアセスメント技法

対訳 ISO 31000:2018（JIS Q 31000:2019）
リスクマネジメントの国際規格［ポケット版］

定価：本体 5,000 円(税別)

2019 年 4 月 15 日　　第 1 版第 1 刷発行

編　　者　一般財団法人 日本規格協会

発 行 者　揖斐　敏夫

発 行 所　一般財団法人 日本規格協会

〒 108-0073　東京都港区三田 3 丁目 13-12　三田 MT ビル
https://www.jsa.or.jp/
振替　00160-2-195146

製　　作　日本規格協会ソリューションズ株式会社

印 刷 所　株式会社ディグ

ISBN978-4-542-40282-9